How to Make a Rolling Machine for Sheet Metal Work

Designed and written by Rob Hitchings

Step-by-step instructions on how to build a hand-operated machine to roll metal sheet and strip into curves and cylinders

T0160729

Practical
ACTION
PUBLISHING

Practical Action Publishing Ltd
27a Albert Street, Rugby, CV21 2SG, Warwickshire, UK
www.practicalactionpublishing.org

© Intermediate Technology Publications 1985

First published 1985\Digitised 2013

ISBN 10: 0 94668 806 0
ISBN 13: 9780946688067
ISBN Library Ebook: 9781780442402
Book DOI: http://dx.doi.org/10.3362/9781780442402

Since 1974, Practical Action Publishing has published and disseminated
books and information in support of international development work
throughout the world. Practical Action Publishing is a trading name
of Practical Action Publishing Ltd (Company Reg. No. 1159018), the
wholly owned publishing company of Practical Action. Practical Action
Publishing trades only in support of its parent charity objectives and any
profits are covenanted back to Practical Action (Charity Reg. No. 247257,
Group VAT Registration No. 880 9924 76).

Acknowledgements

Financial assistance in the development of this machine and in the production of this manual was made available through the Intermediate Technology Development Group from a grant from the Overseas Development Administration. Their assistance is gratefully acknowledged.

Acknowledgement is also given to Rob Hitchings' colleagues at ApT Design and Development, Blockley, Gloucestershire, who assisted in the production of this manual.

Check list

1. Rotate the top roller by hand; check that it turns easily, and that it is not eccentric, bent or damaged.
2. Rotate the handle; check that it also turns easily, and that the two bottom rollers are not eccentric, bent or damaged.
3. Remove the top roller. Check that the two bottom rollers are parallel both horizontally (by looking across the machine) and vertically (by looking at the gap between the rollers from on top of the machine).
4. Check that the handle arm C4 has been welded to crank pin C6 so that it continues in the same direction from spindle B2 as crank box C3. If the arm is attached so that it crosses back over the spindle centre, the user will have much less leverage while rolling metal. (Refer to Fig. 14 for correct layout).
5. Check that there is not excessive play in the connecting bars C1, and that the welds on the cranks C2 and C3 are good.
6. With the cranks C2 vertical, check that they are both parallel and that the cranks C3 are both horizontal, and in line.
7. Check that the hand screws G operate freely.
8. Check that the vertical play in the free ends of the top bearing blocks D is not excessive. Examine all wooden parts for cracks.
9. All moving parts should be oiled.

Contents

Introduction

This manual describes in detail how to construct a simple rolling machine for sheet metal work. The machine is cheap to build and easy to use. It can be made from readily available angle, bar, and hollow steel sections, using basic welding and fabrication techniques. The only equipment essential for its construction is a drilling machine, an electric welder, 'G' clamps and basic hand tools. An angle grinder, mechanical hacksaw and a metal lathe would make it easier, though these tools are not essential.

Construction can be modified to suit locally available materials. The machine can be bolted on to a strong bench or mounted on the stand described.

This machine will be found to be very useful in any small metal workshop to make objects in sheet metal such as stoves, flue pipe, water pipe, buckets, bins, mud-guards, fuel tanks, wind-pump blades, etc.

This rolling machine will prove an invaluable tool in small workshops in the industrialized countries, as well as in the Third World.

Important notice to constructors

The plans and instructions given in this manual *must* be read very carefully at each step of the construction process.

The order described here in which parts are made and assembled is the easiest, and should be followed exactly.

Particular care must be given to the relative positions of parts before they are welded.

Where materials are not available in the sizes specified in the manual, give serious thought as to how your substitution with material of a different size will affect the function of that part of the machine:
— Will the change weaken the finished machine?
— Will it make the machine less durable?
— Will this substitution alter other dimensions given elsewhere in the manual?

Where flat bar or plate of the specified thickness is not available, consider whether you could weld two thinner pieces together around the edges and use this in its place.

It is usually better to use a larger steel section than a smaller one.

Components which slide together or rotate in one another should not be painted on those surfaces, and should be greased or oiled as the machine is assembled. Further oiling from time to time will also prolong the life of the machine.

At the back of this manual is a check list. Please read it both before you build the machine and after you have completed the machine. If all the points listed are O.K. you will be well pleased with your machine.

Uses of the machine

This machine will roll sheet steel up to 16 gauge (approx. 1.5mm thick) x 1 metre wide. It will roll complete cylinders down to 75mm diameter (using the given sizes of rollers).

Some of the shapes and suggested items which this machine can help produce are illustrated below.

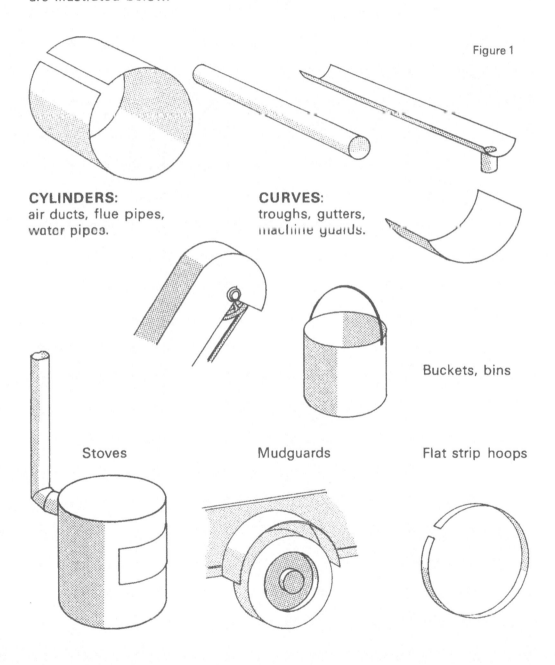

Figure 1

CYLINDERS:
air ducts, flue pipes,
water pipes.

CURVES:
troughs, gutters,
machine guards.

Buckets, bins

Stoves

Mudguards

Flat strip hoops

Description

The machine comprises:

- Three steel rollers, driven by
- Two crank assemblies and handle.
- Two wooden bearing assemblies, mounted on
- The base frame.

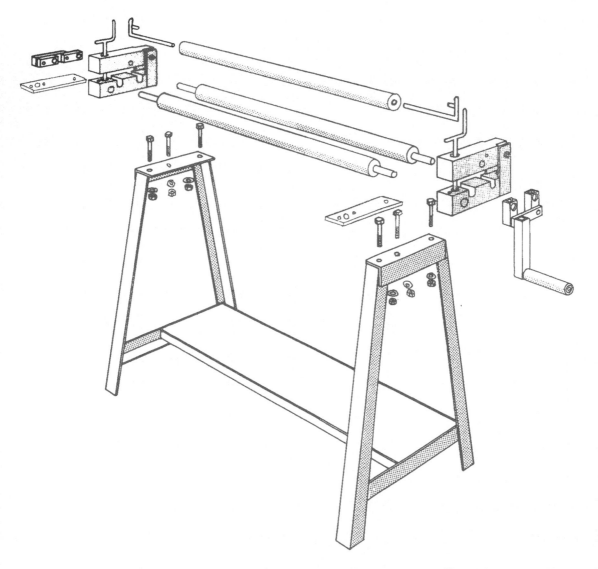

Figure 2

THE ROLLERS

These are made of thick wall tubing. The top roller **A** has a smaller tube welded down the centre to receive the pins **A3**. The bottom rollers **B** have solid spindles **B1** and **B2**. The top roller **A** can be lifted off by withdrawing the pins **A3**. This allows for the removal of a complete cylinder of sheet metal once rolled. (See figures 2 and 6).

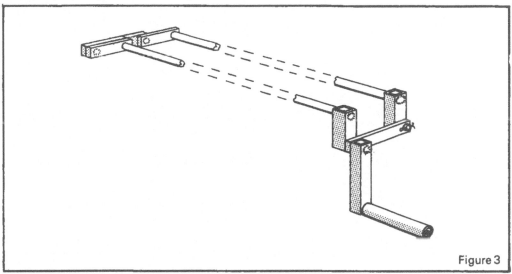

Figure 3

THE CRANK ASSEMBLIES

These link the two bottom rollers, the handle being welded on to the extended connecting rod pin on the front roller spindle.

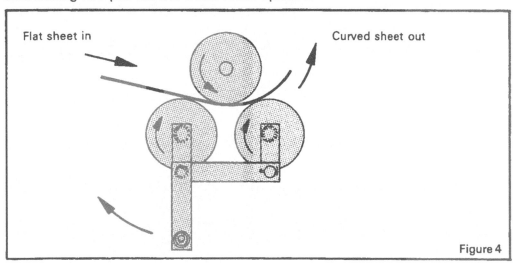

Flat sheet in

Curved sheet out

Figure 4

It is necessary that the two bottom rollers rotate together in order to pull the sheet through as the handle is turned. (Bicycle gears and chains could also be used. See Further Suggestions section.)

THE BEARINGS

The bearing blocks **D** and **E** are made of hard wood (e.g. oak or mahogany). The bracket **F** is steel strip. There is a steel capping strip **E1** which is fitted on top of block **E** for additional strength. When hand screw **G** is turned the top bearing block **D** is raised or lowered, pivoting on bolt **F1**, thus raising or lowering the top roller.

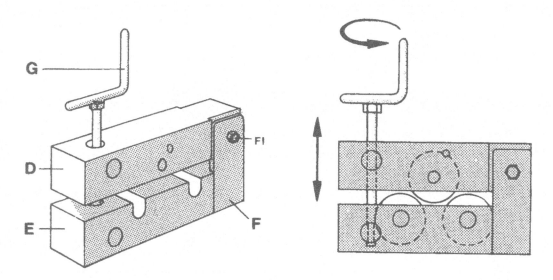

Figure 5

THE BASE FRAME

The machine could be mounted to a bench top. It could have a fold-away frame or simply a rigid frame as described in this manual. The main requirements are rigidity, and suitable height for operation.

Construction

THE ROLLERS ASSEMBLY — PARTS

Part	Name	Quantity	Dimensions (mm)
A	Top roller	1	55 dia. × 1050 steam or galvanised steel pipe
A1	Top roller sleeve	1	20 dia. × 1050 MS tubing
A2	Top roller bushes	4	47 dia. × 5 MS plate
A3	Top roller pin	2	15 dia. × 410 MS bar
A4	Locating pin	2	5 dia. × 10 MS bar
B	Bottom roller	2	55 dia. × 1050 steam or galvanized steel pipe
B1	Driven roller spindle	1	20 dia. × 1200 MS bar
B2	Drive roller spindle	1	20 dia. × 1200 MS bar
B3	Bottom roller bushes	6	47 dia. × 5 MS plate
B4	Washer	4	35 dia. × 2

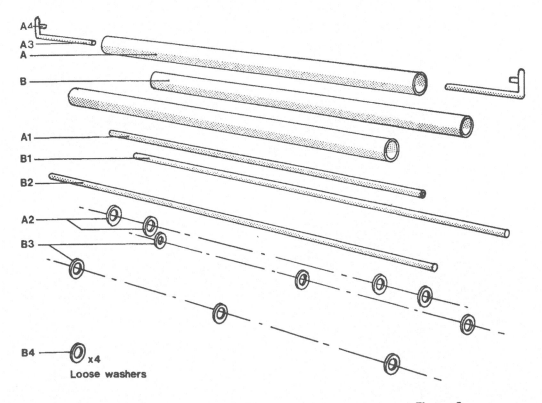

Figure 6

ROLLERS ASSEMBLY

Cut the three rollers to length and file ends square. Cut the tubing for the top roller sleeve **A1** and the round bar for the bottom roller spindles **B1** and **B2**. The top roller bushes can be made either by hand or on a lathe. In both cases first mark out the centres of the four circles on flat strip or sheet (5mm thick). Centre punch and scribe the circles (diameter to suit internal bore of tubing being used). Drill with a small drill e.g. 4mm and enlarge to 20mm (or size to suit outside diameter or pipe being used for **A1**). Cut off the bushes and cut off the corners (Figure 7).

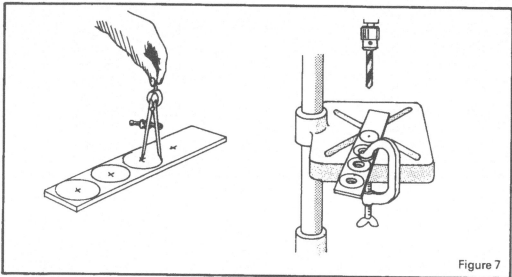

Figure 7

Grind or file the bushes down to the scribed line (or turn in a lathe), until two of them slide into the roller **A**, and two of them are a tight fit. (If turning in a lathe a mandrel will need to be made first.) (Figs. 8 and 9).

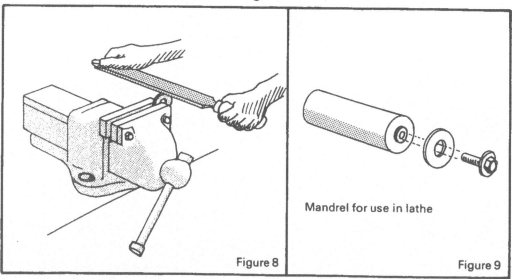

Mandrel for use in lathe

Figure 8

Figure 9

Slide the two slightly easier fit bushes **A2** on to the sleeve **A1** and tack weld in positions shown.

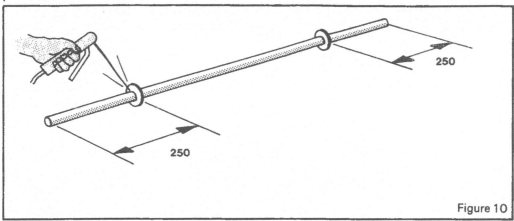

250

250

Figure 10

Slide this assembly in to the top roller **A** and tap in end bushes. Weld to roller and to sleeve.

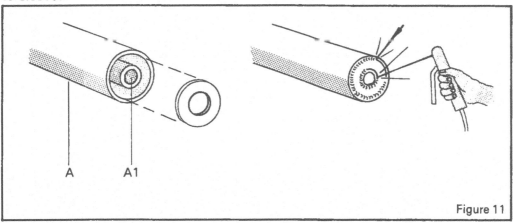

A   A1

Figure 11

Cut pins **A3**, weld the two parts at 90°, making join at 45°. Weld on 5mm dia. locating pins in position shown (Fig.12). The clearance hole in the wood-block bearing should be 8.5mm dia. × 15mm deep.

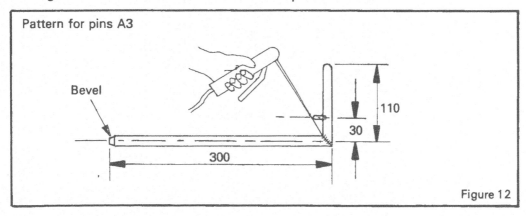

Pattern for pins A3

Bevel

110

30

300

Figure 12

Next mark out the bushes for the bottom rollers. Drill holes 20mm or to suit spindle diameter. Cut and file or turn as before. Position one bush on each shaft halfway, and tack weld in place. Insert the spindles into the rollers and tap in the end bushes. Take care to position the spindles accurately, then weld in place.

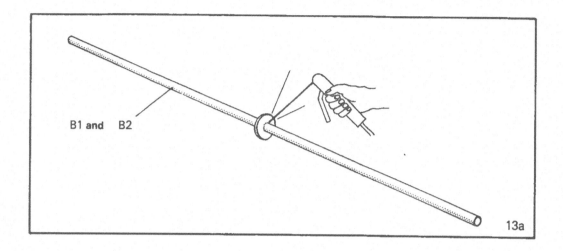

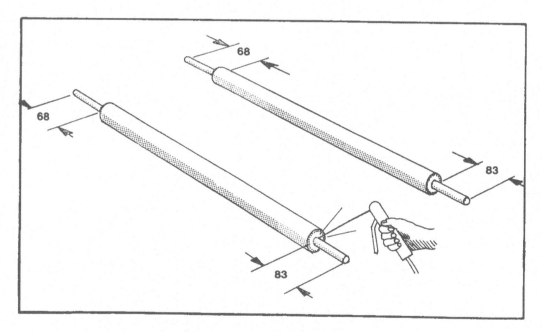

Figure 13

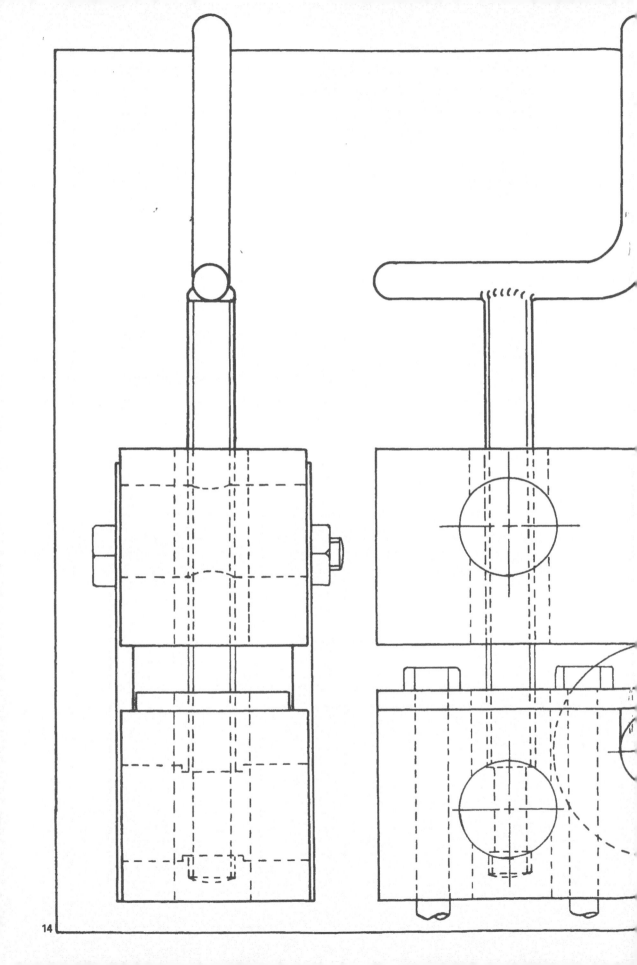

14

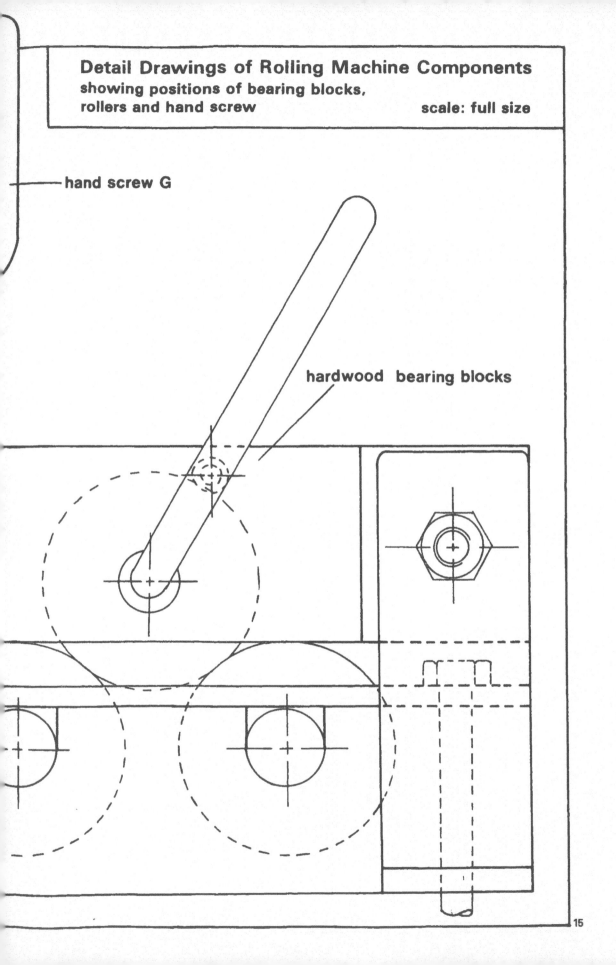

Detail Drawings of Rolling Machine Components
showing positions of bearing blocks, rollers and hand screw

scale: full size

hand screw G

hardwood bearing blocks

15

THE CRANK ASSEMBLIES —
PARTS

Part	Name	Quantity	Dimensions (mm)
C1	Connecting bar	2	12 × 25 × 95 MS bar
C2	Crank	2	12 × 25 × 65 MS plate
C3	Crank (box)	2	25 × 25 × 65 MS square tube
C4	Handle arm	1	25 × 25 × 200 MS square tube
C5	Crank pin	3	15 dia. × 32 MS bar
C6	Crank pin	1	15 dia. × 70 MS bar
C7	Washer	3	25 dia. × 1.5
C8	Split pin	3	3 dia. × 25
C9	Handle spindle	1	15 dia. × 175 MS bar
C10	Handle sleeve	1	20 dia. × 138 MS tube
C11	Handle	1	25 dia. × 150 MS tube
C12	Handle bush	1	20 dia. × 15 MS tube

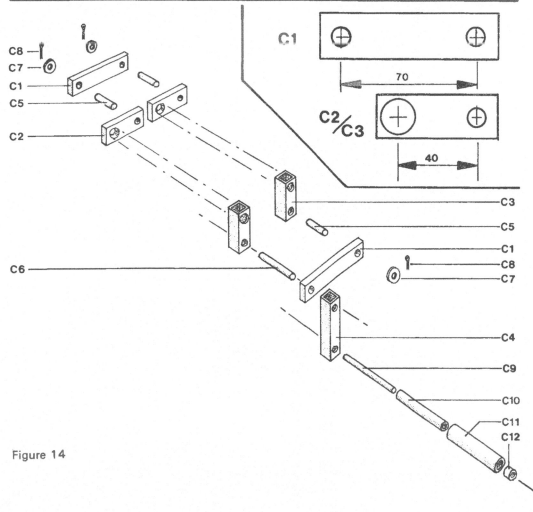

Figure 14

CRANK AND HANDLE ASSEMBLY

Cut, mark and drill accurately connecting bars **C1** and cranks **C2** and **C3**. Cut three pins **C5** and one **C6**. Weld the pins **C5** and **C6** into their respective cranks.

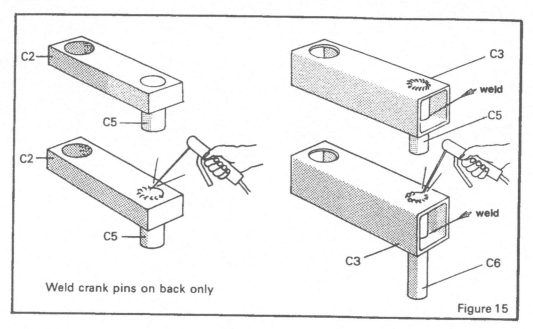

Weld crank pins on back only

Figure 15

Take two washers **B4** and slip them on to the short end of the bottom roller spindles. Push on the cranks **C2** and weld as indicated.
Similarly slide on washers **B4** and cranks **C3** to the longer end.
DO NOT WELD YET.

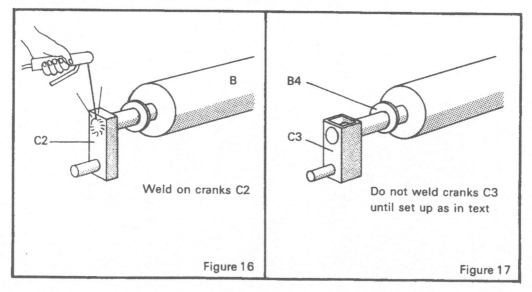

Weld on cranks C2

Figure 16

Do not weld cranks C3 until set up as in text

Figure 17

SETTING UP CRANKS C3

This requires as much care and accuracy as possible. So read carefully, and double check before welding.

The cranks **C3** need to be at right angles to the cranks **C2**. This shortened illustration shows this clearly (Figure 18).

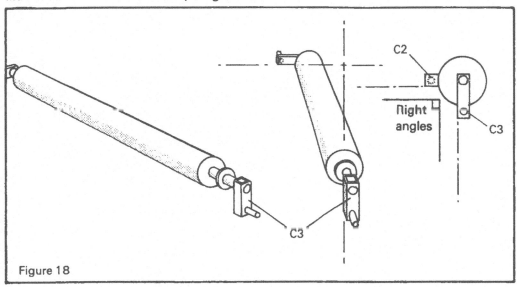

Figure 18

Here are two methods which could be used to set this up:

Method 1. Put the rollers **B** in a vice. Set crank **C2** horizontal with a spirit level. Then using a square with the spirit level, set the crank **C3** vertical. Tack weld and check before welding fully.

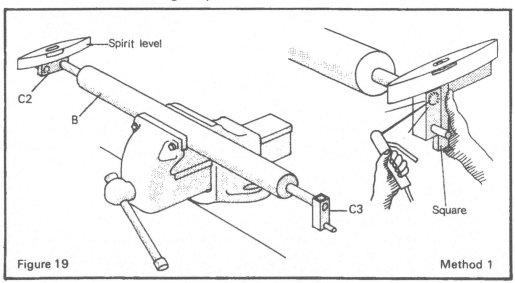

Figure 19   Method 1

Method 2. Have someone hold a square on **C2** while you hold a straight metal bar against **C3**. Check with one eye the alignment of the metal bar with the square. Tack weld first, check the alignment, then weld fully. The box section cranks **C3** can be welded inside and on the ends as shown (Figure 21).

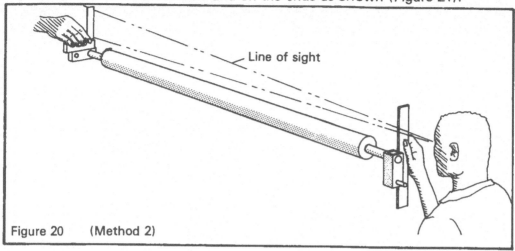

Line of sight

Figure 20   (Method 2)

THE HANDLE

Cut pieces **C4, C9, C10, C11** and **C12**. Mark out and drill **C4**. Weld the spindle **C9** in to the handle arm **C4** (Figure 22i). Slide tube **C11** over **C10** with 3 mm. protruding, and weld as shown (Figure 22ii). Slide **C10/11** onto the spindle followed by bush **C12**. Weld the end of bush **C12** to the end of the spindle **C9** (Figure 22iii). Take care **not** to weld on to **C11** as this should rotate freely on the spindle.

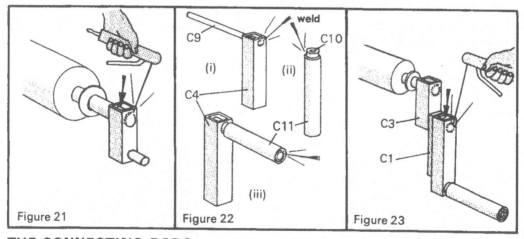

Figure 21

weld

C9   C10

(i)   (ii)

C4

C11

(iii)

Figure 22

C3

C1

Figure 23

THE CONNECTING RODS

The connecting rods **C1** should be an easy slide fit on the pivot pins. File out carefully if necessary. Slide one connecting rod on to the longer pin **C6** followed by the handle assembly, allowing 2 - 3mm side play. Line up the handle with the crank **C3** and weld as shown (Figure 23).

THE BEARING BLOCK ASSEMBLIES — PARTS

Part	Name	Quantity	Dimensions (mm)
D	Top bearing block	2	50 × 50 × 210 hardwood
E	Bottom bearing block	2	50 × 50 × 210 hardwood
E1	Capping plate	2	40 × 5 × 210 MS strip
F	Mounting bracket	2	40 × 5 × 280 MS plate
F1	Pivot bolt	2	12 dia. × 70
G	Hand screw	2	10 dia. × 140 MS bar
G1	Threaded spindle	2	12 dia. × 150 studding
G2	Threaded pin	2	25 dia. × 50 MS bar
G3	Drilled pin	2	25 dia. × 50 MS bar
G4	Sleeve	2	12 dia. × 8 MS tube

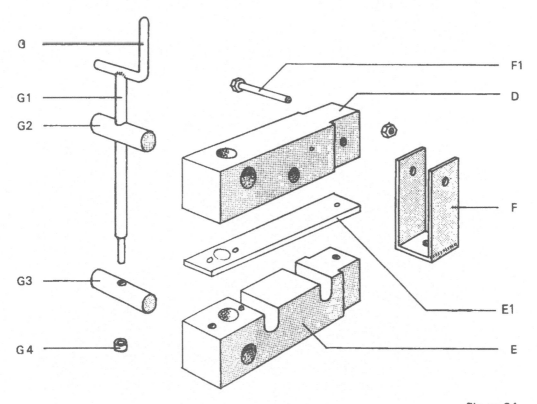

Figure 24

BEARING BLOCK ASSEMBLY

Make two of each part shown opposite. The following describes the fabrication of one assembly. Mark out and drill capping plates **E1**. Mark out and cut hardwood timber blocks **D** and **E**. Drill holes shown and cut rebates on sides of both blocks, and bottom of block **E**. The two grooves in **E** can be made by first drilling a 20mm diameter hole, and then cutting down with a saw. (For overall dimensions see list above.) Check that the roller shafts fit the grooves.

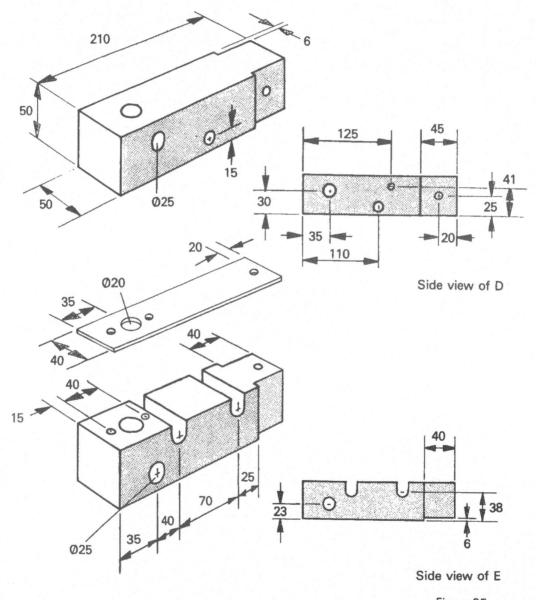

Side view of D

Side view of E

Figure 25

Bracket **F** could be made from one piece of steel heated up and bent, then drilled. It may be easier and more accurate to make it in three pieces. Cut and weld as illustrated (Fig. 26). Y = 84mm for 55 dia. rollers, Y = 90mm for 60 dia. rollers. Clamp on to the wood blocks, with bolt **F1** in place whilst tack welding. Unclamp and weld inside and out.

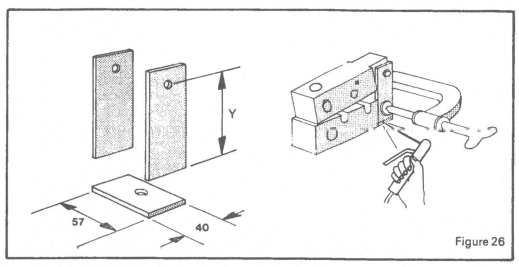

Figure 26

The pins **G2** and **G3** should be cut next. These should slide easily into the holes in the bearing blocks. Mark the centres, punch and drill accurately with a 3mm drill (Figure 27). Enlarge the holes in **G2 (2 only)** to tapping size for the 10mm or 12mm threaded spindle. Tap (cut) the threads.

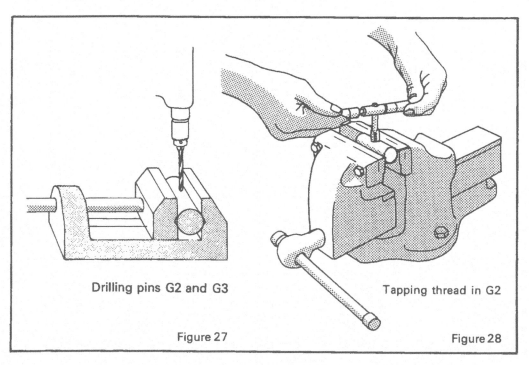

Drilling pins G2 and G3

Tapping thread in G2

Figure 27

Figure 28

Enlarge the holes in **G3 (2 only)** to 1 mm **less** than the tapping size. Take the threaded spindle **G1** and turn or file the end to fit the hole in **G3**. If filing, make it as round and smooth as possible, and with a neat sharp shoulder (Figure 29).

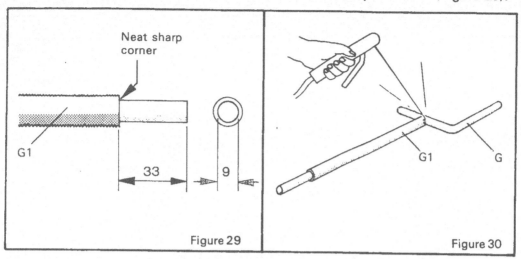

Figure 29

Figure 30

Cut and bend the hand screw **G** and weld on to the spindle **G1**. Slide pin **G2** in to its hole in block **D**, and pin **G3** in to block **E**. Screw the spindle down through block **D** and insert the turned end in to its hole in pin **G3** (Figure 31 (i)).

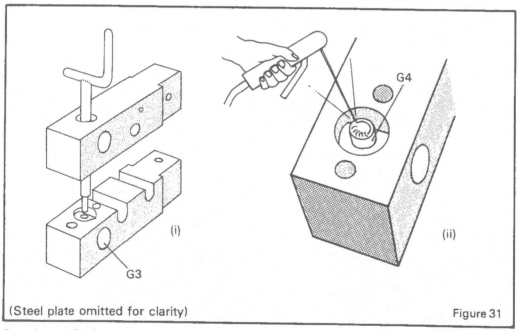

(Steel plate omitted for clarity)

Figure 31

Cut sleeve **G4** from pipe (or drill a piece of bar to suit). Turn the assembly upside down and position sleeve **G4** on to the protruding end of the spindle **G1**. Carefully weld the ends together (Figure 31 (ii)).

Check that the spindle rotates freely. Apply oil.

BASE FRAME ASSEMBLY —
PARTS

Part	Name	Quantity	Dimensions (mm)
H1	Mounting plate	2	50 × 50 × 210 MS angle
H2	Legs	4	40 × 40 × 850 MS angle
H3	Tie	2	40 × 40 × 600 MS angle
H4	Shelf support	2	40 × 40 × 1150 MS angle
H5	Cross brace	2	20 × 20 × 1300 MS angle
H6	Bolt, nut and washer	6	10 dia. × 75
H7	Shelf	1	350 × 12 × 1150 wood

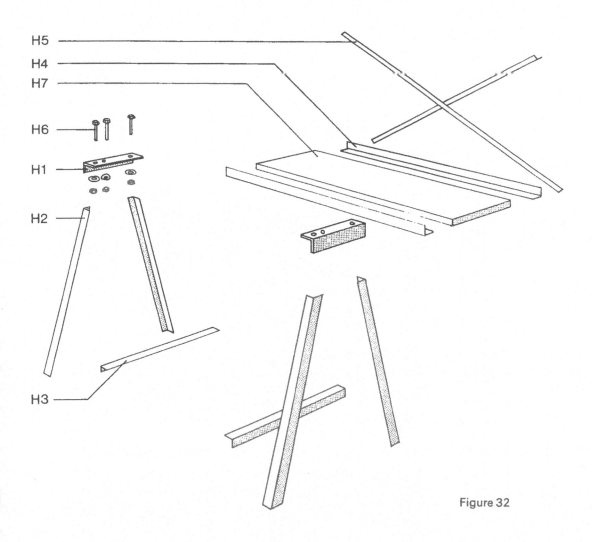

Figure 32

Cut mounting plates **H1** and drill as shown (Fig. 33). Cut legs **H2** and ties **H3**. Lay out the legs and weld as shown (Fig. 34). Assemble the machine **before** welding the rest of the base frame.

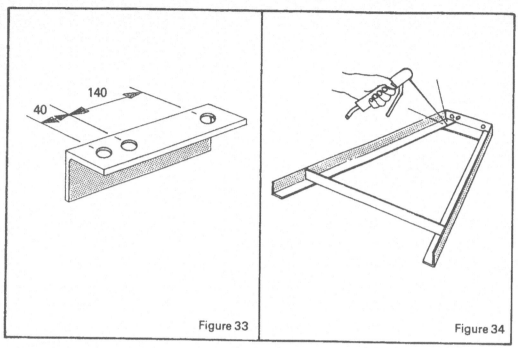

Figure 33	Figure 34

Assembly

Using bolts **H6** loosely bolt on the bearing block assemblies to the mounting plates **H1**. Note left and right hand. Note also that the capping plates **E1** are held on top of the bearing blocks **E** by bolts **H6**. Stand up the legs and clamp to some suitable support. Insert the bottom rollers at the same time inserting the crank pin **C5** in to the connecting bar **C1**. The washers **B4** go outside the blocks. Check that the rollers will rotate freely together. If not, the grooved bearings in the right hand block **E** may need to be filed out slightly. Fit on the other connecting bar, washers and split pins. (Holes should be drilled to allow some clearance.) Check for free rotation on the left-hand bearing block **E**.

Remove the rear bolts **H6** and insert the mounting brackets **F** and bolts **F1**. Replace bolts **H6** and tighten all four.

The rest of the base frame can now be welded up, **H4** and **H5**. (Check the frame for squareness before welding.) The shelf **H7** could be wood or metal.

The top roller can now be added. Use plenty of oil on pins **A3** and on the wood bearings in block **E**.

The machine should now operate.

Operation

To set up the machine for rolling sheet metal first turn the two hand screws **G** clockwise to raise the top roller until the sheet slides right through (Figure 35). Remove the sheet metal and check that the top roller is the same height both sides. View this from the front.

Turn both hand screws half a turn anti-clockwise to lower the top roller. Push the sheet metal in and at the same time turn the main handle clockwise. It should easily grip the sheet and pull it through. Stop when the end of the sheet is just past the top of the front roller and reverse the rotation (Figure 36).

Stop again at the beginning and increase the pressure on the hand screws (turn anti-clockwise half a turn). Repeat this process until the required curve is achieved. The amount of increase curve added each time varies according to type of metal and thickness being rolled. If you add too much pressure at once it will become difficult to turn the handle.

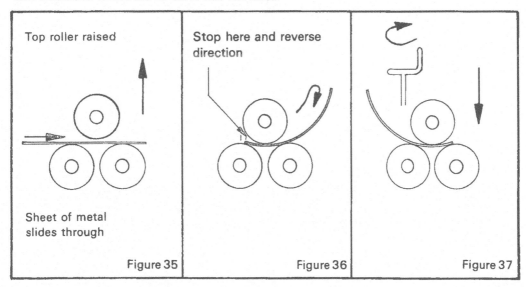

Top roller raised

Sheet of metal slides through

Figure 35

Stop here and reverse direction

Figure 36

Figure 37

ROLLING CYLINDERS

When rolling cylinders it is best to cut the sheet metal to the correct length first. You may require an overlap for riveting, so remember to add this on. The way to find the length required is to multiply the diameter by $\frac{22}{7}$ (3.142).

For example:

A cylinder of 250mm diameter would require:
250 × 3.142 = 785.5mm length of metal sheet,
plus 20mm for overlap = 805.5mm

Further suggestions

CHAIN DRIVE

Where the facilities and materials are available the bottom rollers could be connected by chain drive (Fig. 38). In this case the shafts could be short at the other end. This end would need the same lengths, but the cranks would be replaced by the gear wheels. Ones with 18 teeth are ideal. The handle arm would be 50mm longer and welded directly on to the roller spindle.

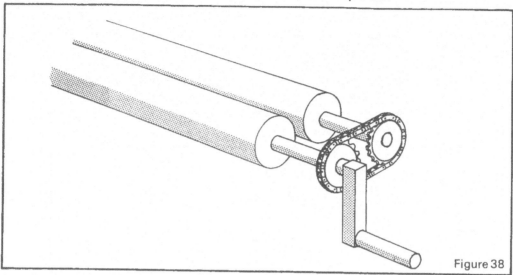

Figure 38

FOOT REST

It may be found helpful to add a bar or plate to stand on. This could be a straight tube or angle connecting the front two feet, or could be a fold-down plate. (See Figure 39).

Figure 39

ROLLING ROUND BAR

Round bar up to 8mm diameter can be rolled more easily if grooves are cut in the two bottom rollers (figure 40). This will weaken the tubing unless very thick wall tube has been used. This could be overcome by turning a plug of solid bar or smaller pipe to fit tightly down the roller for 100mm or so (Figure 41). This plug should be welded at the end, and also at four holes drilled through the roller at **Z**. The holes can be filled in with weld and filed smooth.

The groove could be either cut on a lathe or by one person grinding with an angle grinder as the other person turns the handle. **Eye protection should be worn by both persons.**

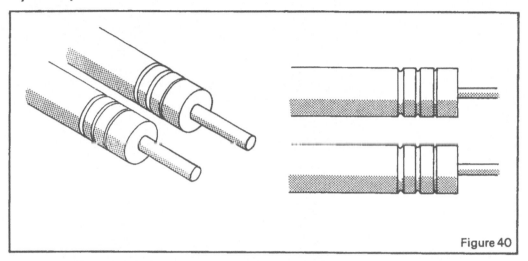

Figure 40

The grooves should ideally be half rounds of the sizes of wire and bar that you may wish to roll. Three, five and six millimetres would be useful, or three 'V' grooves which allow for some variations. The grooves on the front and back rollers must be in line.

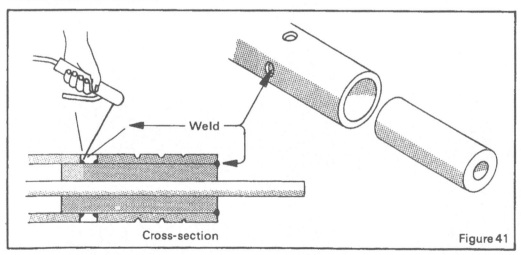

Weld

Cross-section

Figure 41

Milton Keynes UK
Ingram Content Group UK Ltd.
UKHW011338070824
1191UKWH00034B/254